设计有约 2（下册）

inSIDE deSign

香港方黄建筑师事务所住宅空间设计系列专集　方峻 著

华中科技大学出版社
http://www.hustp.com
中国·武汉

i Fong, a multi and trans-industry designer

in Hong Kong, China. studied the bachelor/postgraduate/doctor

programs and degrees on philosophy, construction engineering and

design management in Inter American University,

Politecnico di Milano, Huaqiao University, and Hong Kong

Polytechnic University. He is a Professional Associations of Hong Kong

Interior Design Association (PM00408),

Member of International Federation of Interior Architects/ Interior Designers (0281),

Institute of Interior Design of Architectural Society of China (8030),

Lighting designer of register of China.

The works not only was awarded the golden award

at Indoor Design Contest for the 1st International Building Landscape,

also was awarded by awards and special honors

at indoor design contests home and abroad.

Be more have been inserted by American *Indoor Design* and

various well-known magazines.

Here are published and highly acclaimed industry of

personal album design have been

Inspiration Design,

Inside Design I, Inside Design II…Inside Design V,

Mangement System and Application for Installation Art Project etc.

方峻（TFong）

建筑空间与多元跨界的中国香港设计师。

先后在美国美联大学、意大利米兰理工学院、香港理工大学、国立华侨大学

接受哲学、建筑设计、设计管理的学士／硕士／博士等教育，

同时也是香港室内设计协会专业会员、国际室内建筑师设计联盟会员、中国建筑学会室内设计分会会员、中国注册照明设计师。

其作品不但荣获首届国际建筑景观室内设计大奖赛金奖，

还获得过多项国内外室内设计的奖项与荣誉；更被美国《室内设计》和

各类知名专业杂志数次刊载。相继出版且备受业界好评的

个人设计专辑分别有《"悟"设计》《设计有约 1》《设计有约 2》……《设计有约 5》

《装置艺术项目管理体系与应用》。

Contents
目录

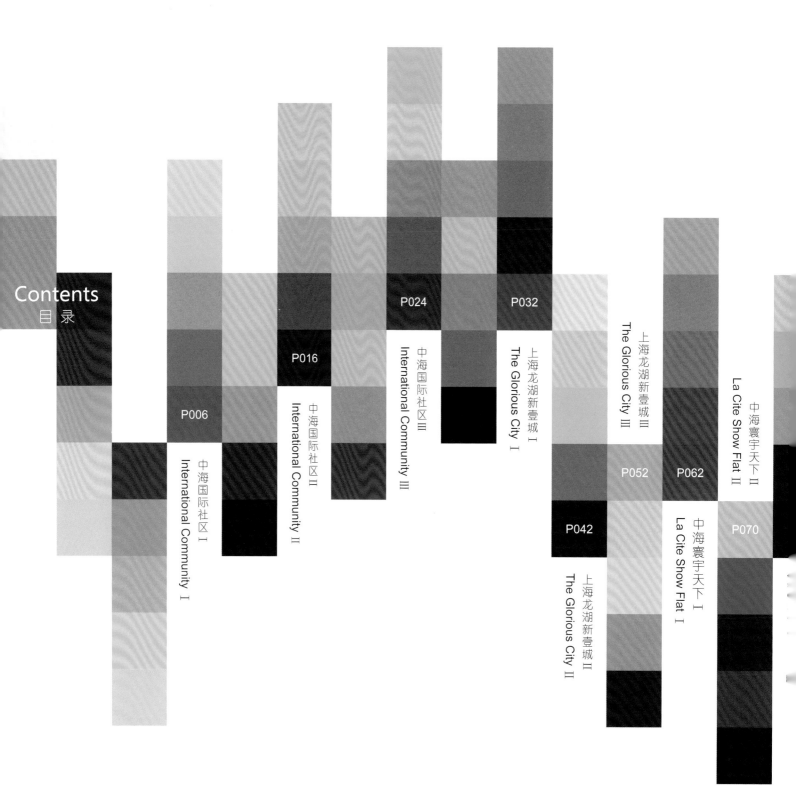

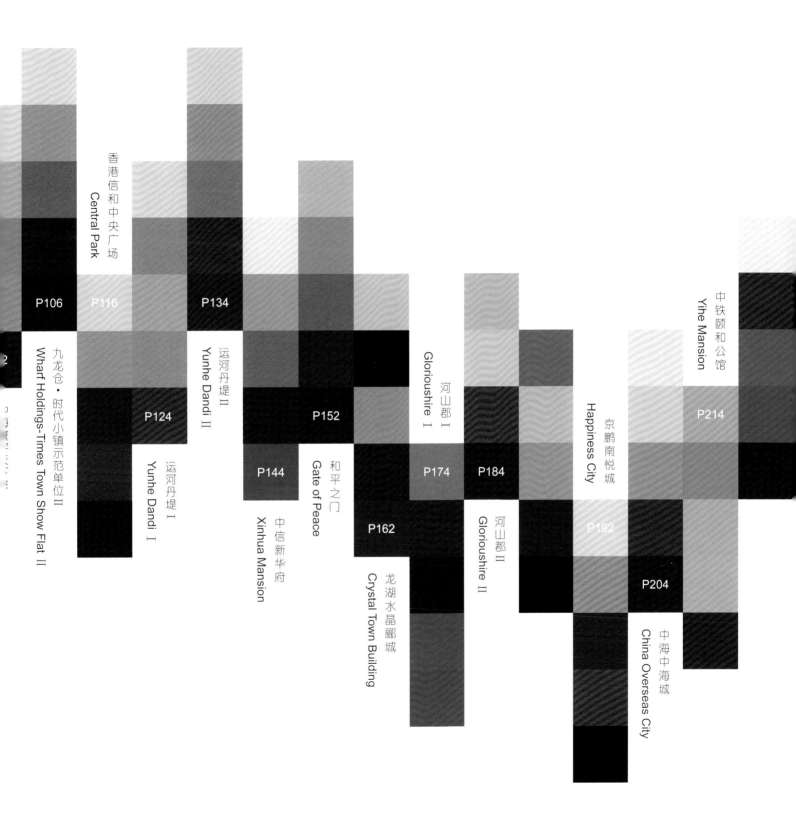

International Community Ⅰ

中海国际社区Ⅰ

International Community Ⅱ

中海国际社区Ⅱ

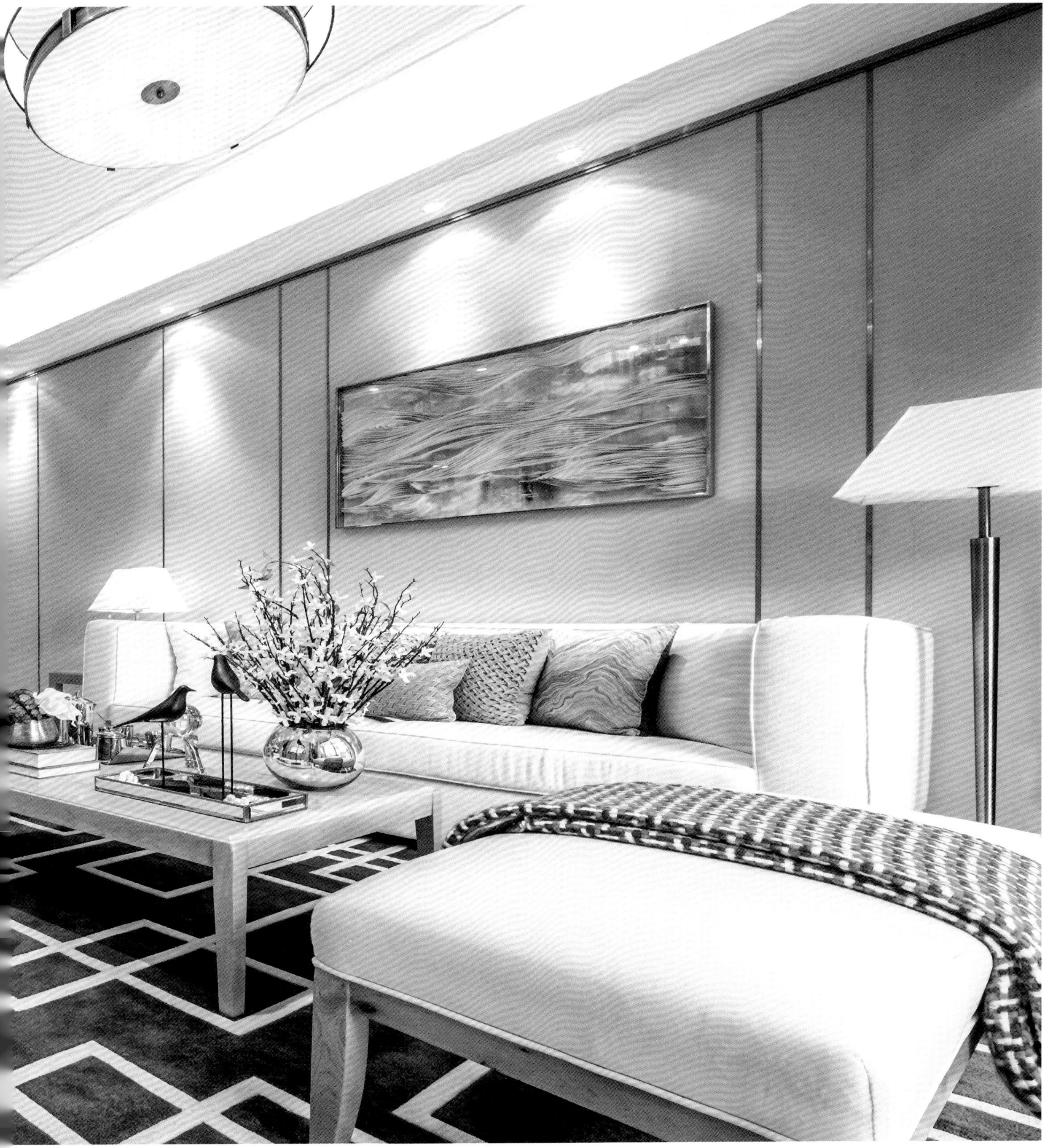

International Community III

中海国际社区III

The Glorious City Ⅰ

上海龙湖新壹城Ⅰ

The Glorious City Ⅱ

上海龙湖新壹城Ⅱ

The Glorious City Ⅲ

上海龙湖新壹城Ⅲ

La Cite Show Flat 1

中海寰宇天下 1

Interio Design – Kinney Chan & Associates – FF&E – Photo – FongWong

La Cite Show Flat II

中海寰宇天下 II

Interio Design – Kinney Chan & Associates – FF&E – Photo – FongWong

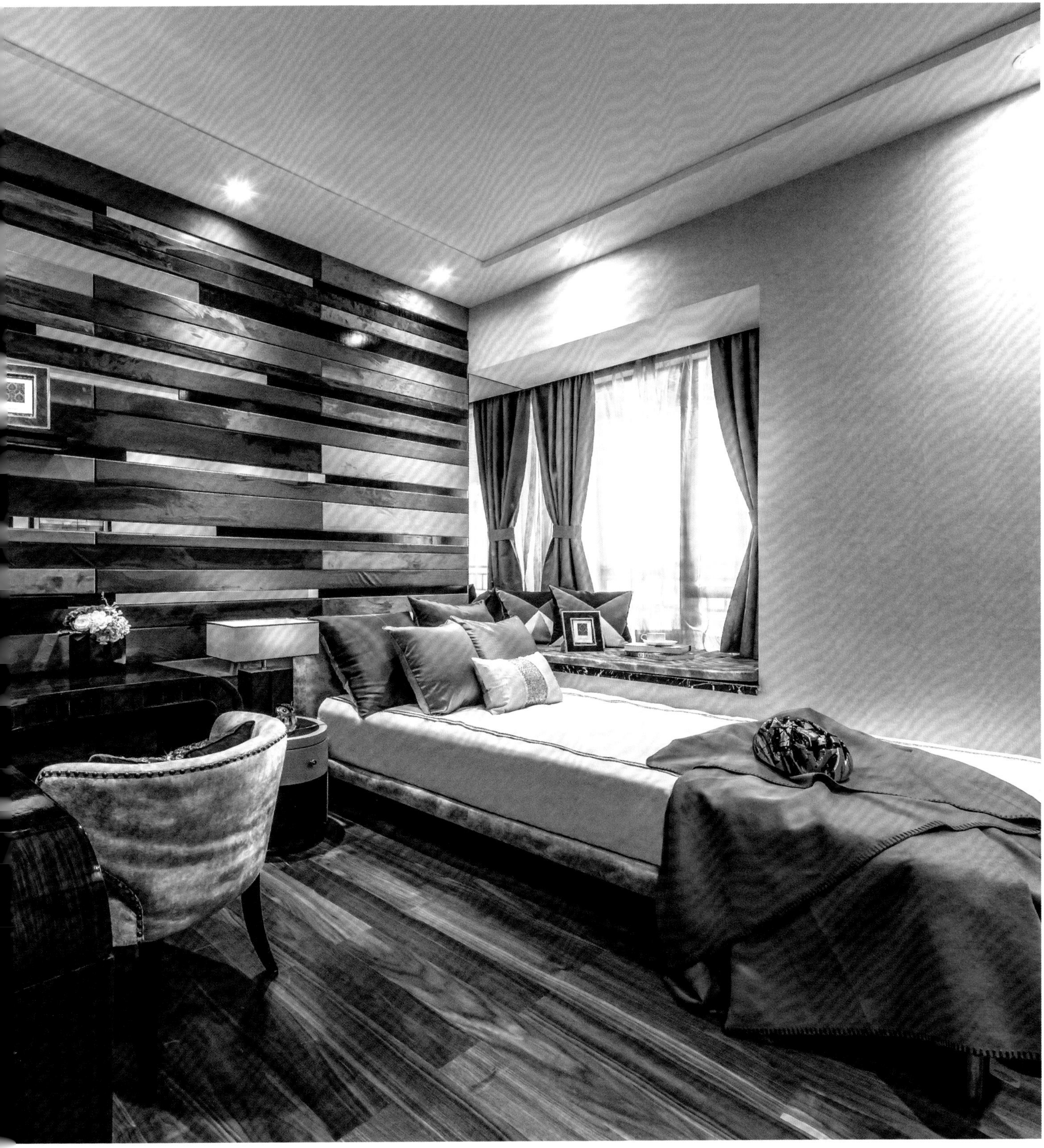

La Cite Show Flat III

中海豪宅天下III

Interio Design – Kinney Chan & Associates – FF&E – Photo – FongWong

La Cite Show Flat Ⅳ

中海寰宇天下Ⅳ

Interio Design – Kinney Chan & Associates – FF&E – Photo – FongWong

Wharf Holdings-Times Town Show Flat Ⅱ

九龙仓·时代小镇示范单位Ⅱ

It is the color of autumn – modest and soft- bringing people to the natural landscape in sentimental nostalgia free of brusqueness and ostentation. It is the past autumn with memorable people and obsession that tarnish in tranquility and mellowness.

榛果黄是属于秋天的色彩，低调柔和，没有丝毫的突兀与张扬，这种黄色让人宛如置身于大自然这幅风景画中，带着些许怀旧的情绪，旧人旧物，那一年的秋色，予人恬淡、宁静又让人不禁浮想联翩。

Central Park

香港信和中央广场

Lively and elegant pansy. The pansy is also named butterfly flower, and is rather popular in Europe for its colorfulness. Some white, some yellow, and some purple of the pansy create a lively picture. Refreshing white, brisk yellow, mysterious purple, all mix harmoniously with a stratified, pleasant and luxurious feeling just appropriately interpreting the notion of longing.

三色堇，生动雅致。三色堇又名蝴蝶花，色彩斑斓，在欧美颇受欢迎。三色堇的色调跳跃生动，白色纯净清爽、黄色愉悦轻松、紫色神秘优雅。这三种色彩的搭配，相得益彰，很有层次感，营造出愉快和华丽的氛围。作为表现"思慕"情怀的花卉，三色堇充满生机，情趣盎然。

Yunhe Dandi Ⅰ

运河丹堤Ⅰ

Xinhua Mansion

中信新华府

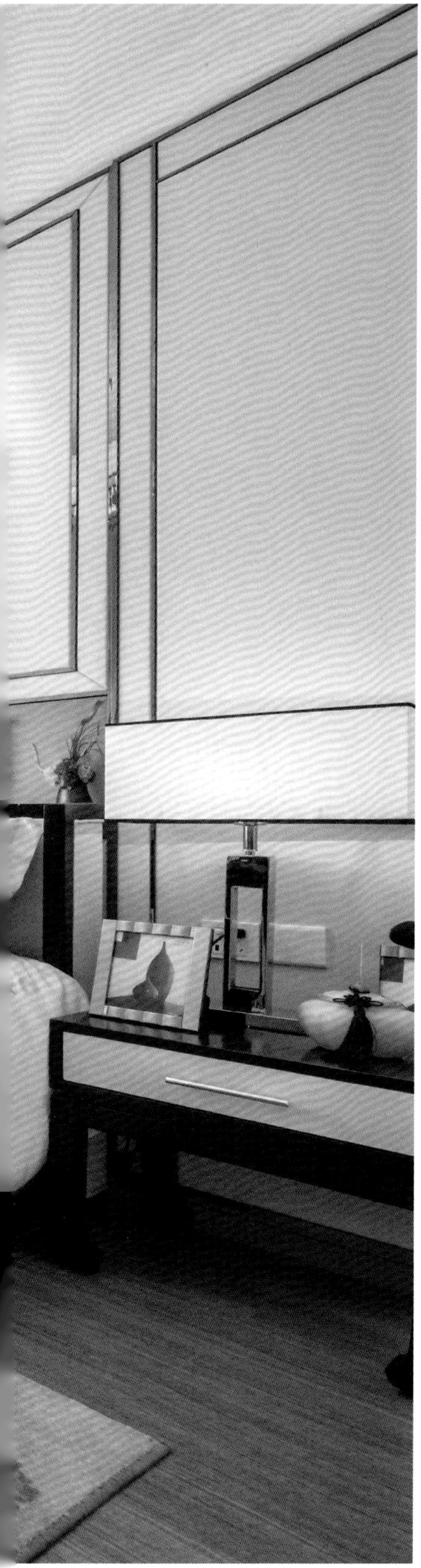

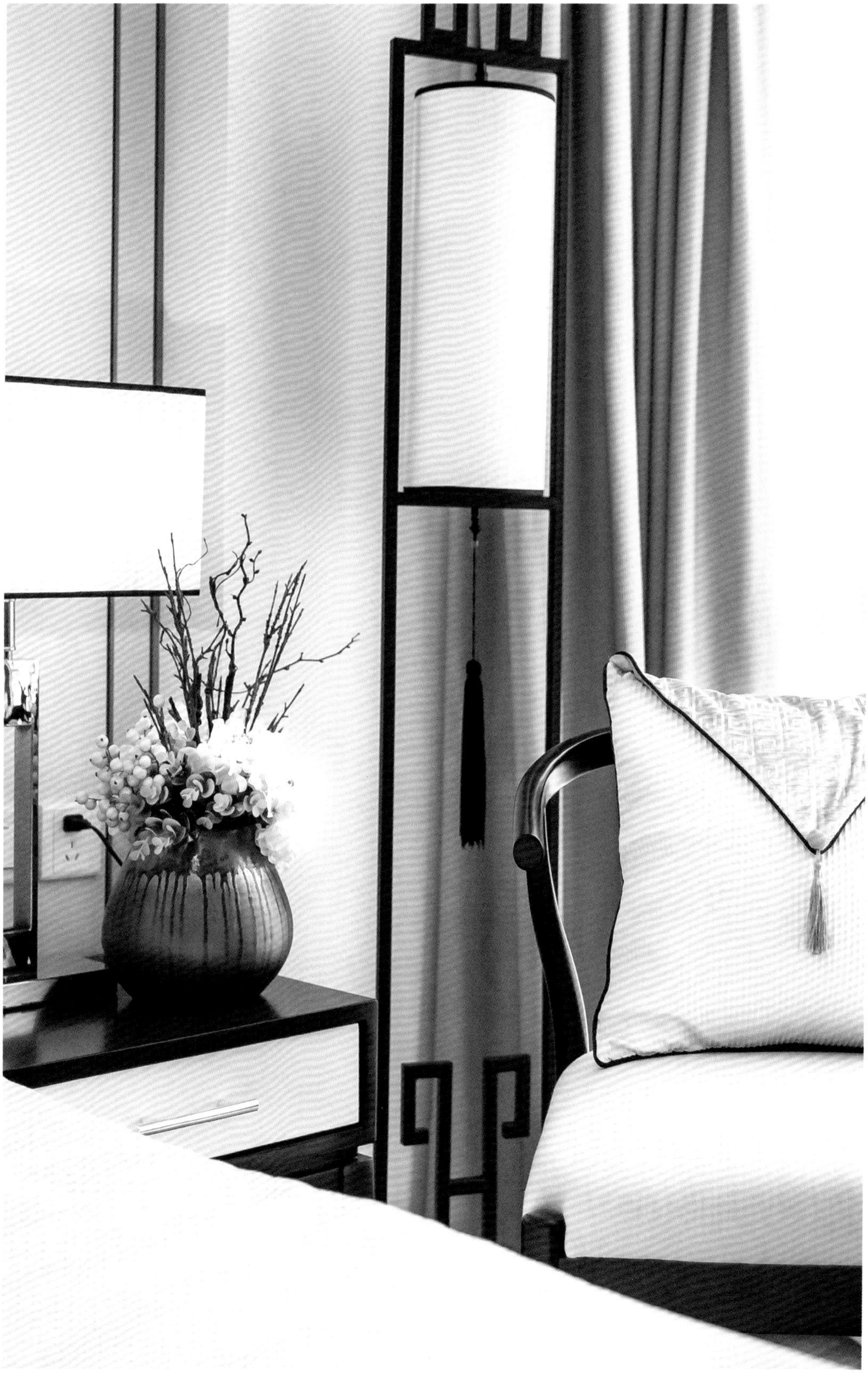

Gate of Peace

和平之门

Quietly elegant and peaceful: Plumbago Auriculata As long as there is sunshine, it can bloom into a cluster of beautiful blue. The blue color of Plumbago Auriculata, like its name, is gentle and beautiful, low-key and soft, impressing people as being tranquil and reserved.

蓝雪花，淡雅幽静。只要有阳光，就能开出一床很美很美的幽蓝。蓝雪花的色彩就如其名一样，轻盈美丽，低调柔和，给人一种清静自持的印象。

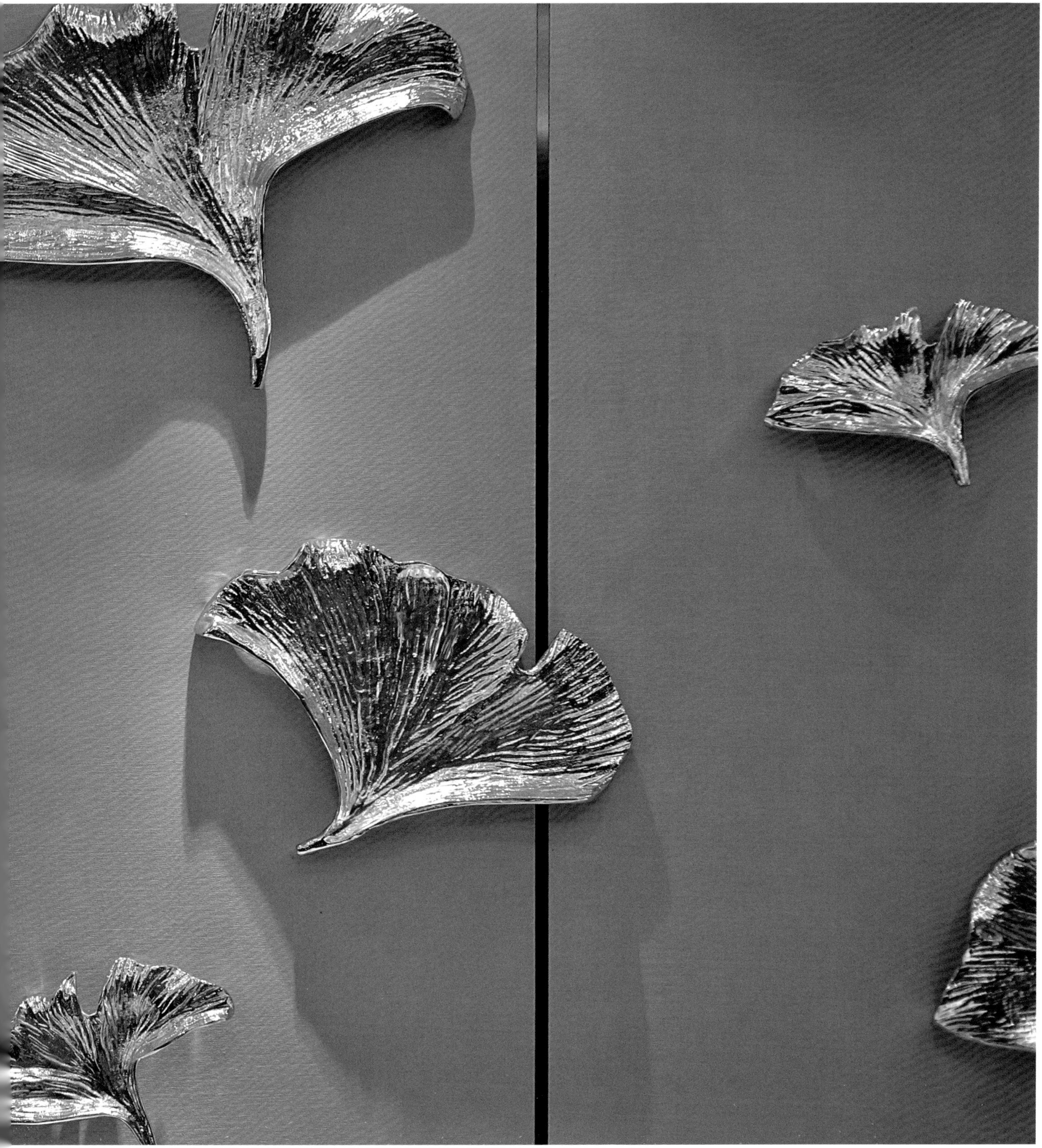

Crystal Town Building

龙湖水晶郦城

Noble and calm. The tawny petals of Day Lily remind people of the color of the earth and make people feel peaceful and stable. This color can easily match up with other colors, representing delicate and exquisite quality.

萱草花，高贵沉稳。萱草花褐色的花瓣令人联想到土地的色彩，给人安定踏实的感觉。这种颜色很容易与其他颜色搭配，呈现出精致考究的气质。

Glorioushire Ⅰ

河山郡Ⅰ

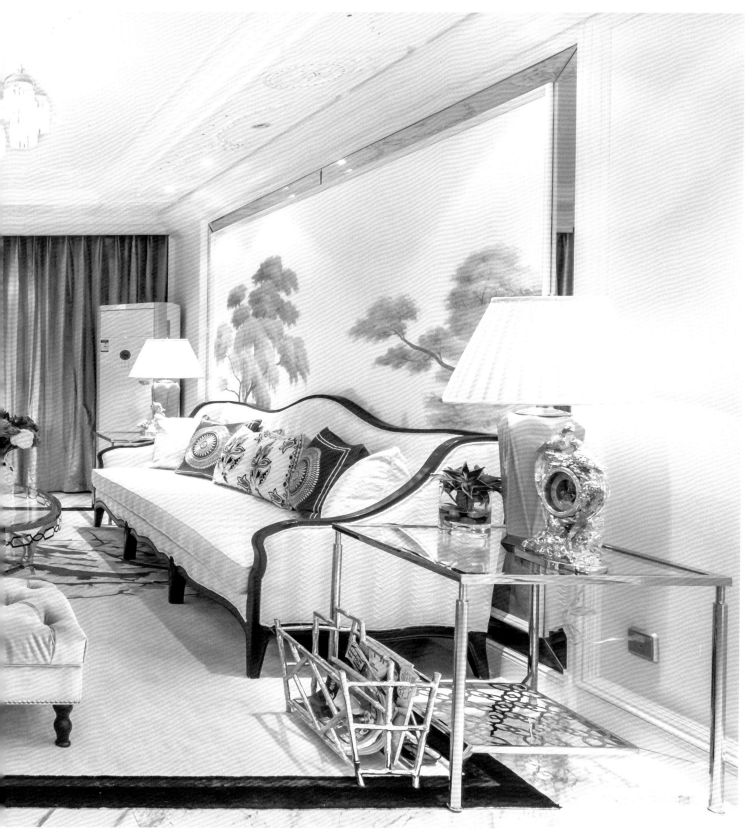

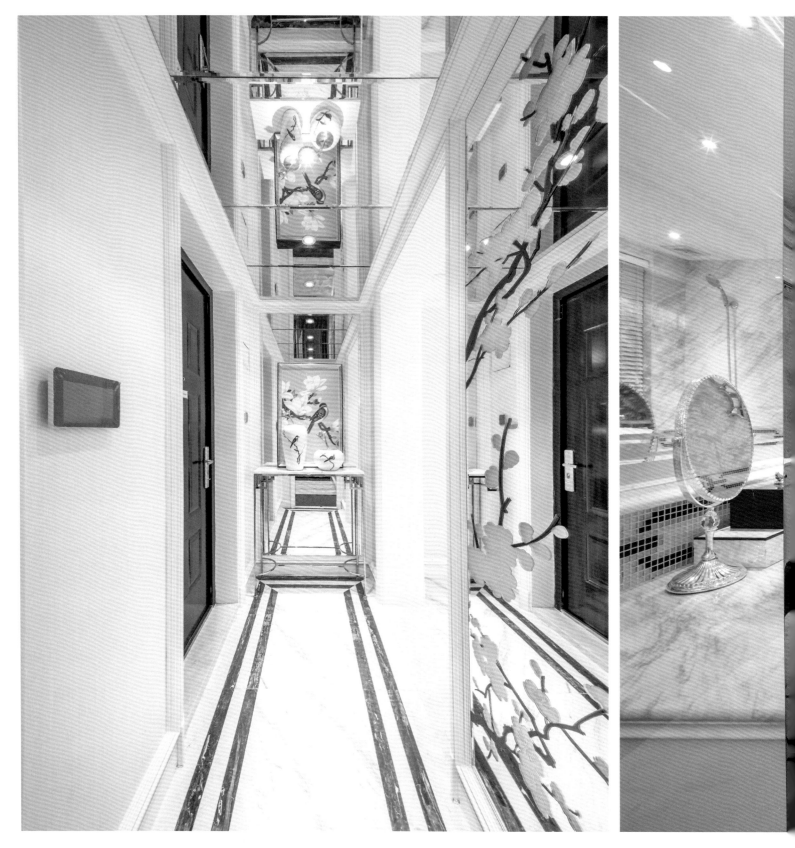

Glorioushire Ⅱ

河山郡Ⅱ

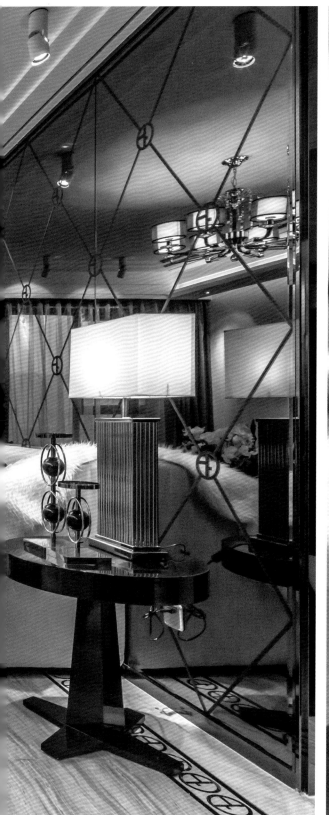

Happiness City

京鹏南悦城

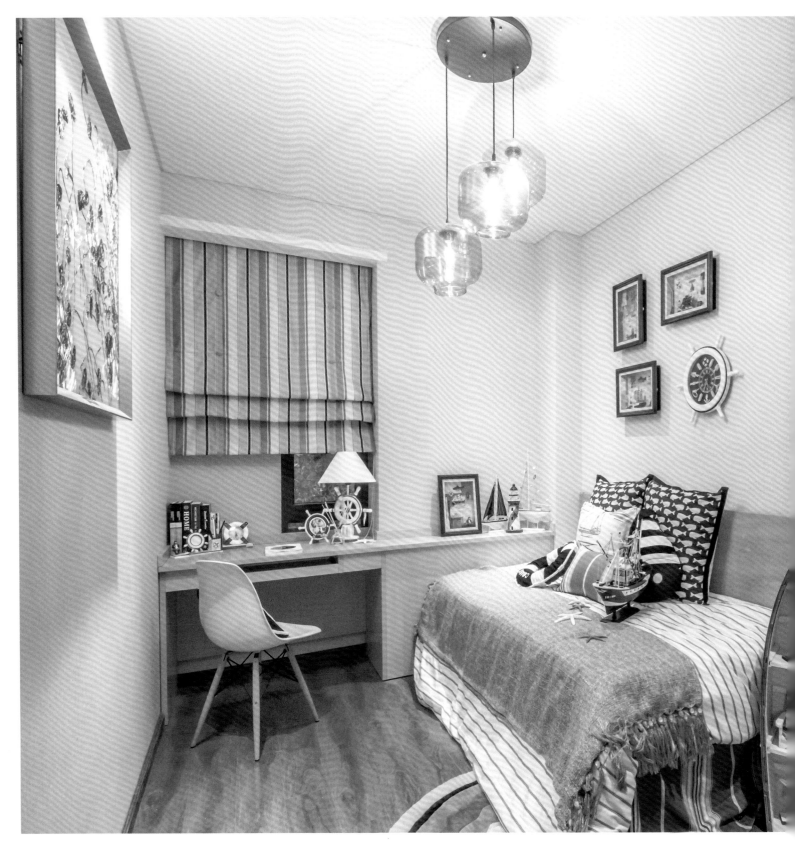

China Overseas City

中海中海城

Yihe Mansion

中铁颐和公馆

Acknowledgments 鸣谢

本书得以顺利面世，全赖各方的参与与支持，在此由衷感谢"香港方黄建筑师事务所"全体同仁的努力与付出！也衷心感谢以下合作伙伴对我司的大力支持和信任！

九龙仓（中国香港）	信和（中国香港）	新榕建筑置业（中国澳门）
置地集团（中国香港）	南益（中国香港）	新建利建筑置业（中国澳门）
中海地产	龙湖地产	碧桂园地产
华侨城地产	绿地集团	中铁集团
华润地产	珠江地产	中粮地产
远航地产	远大地产	世茂地产
中能置业	大华地产	长航地产
滕王阁地产	光达地产	恒河地产
华人地产	新希望地产	恒丰地产

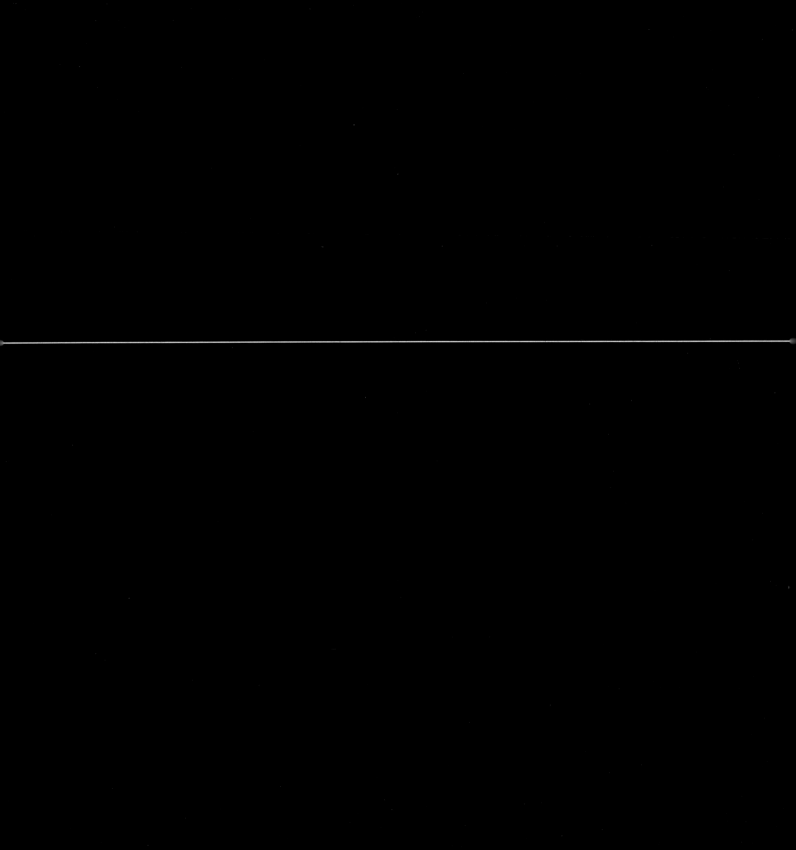

居住、商业空间整体设计与统筹管理的营运机构。事务所经过近二十年的发展，从香港到深圳、上海、成都相继成立了高端专业的国际化设计与管理服务团队，团队至今拥有中国、意大利、加拿大等多个地区的优秀合伙人与设计师。事务所以不断创新务实的设计理念及服务赢得了众多客户的支持与信赖。

合作客户既有香港、澳门等地的知名房地产开发商，如九龙仓、信和、置地、南益地产、新建利建筑置业……也有内地大型房地产开发企业，如中海、龙湖、碧桂园、世茂以及中粮、华润、绿地、中铁、华侨城地产等。项目作品已遍布纽约、香港、澳门、北京、上海、广州、深圳、厦门、天津、成都、重庆、沈阳等多个城市及地区。

Hong Kong Fong & Wong Architects & Associates was established in Hong Kong in 1997, which specializes in integrated design and management of diversified residential and commercial space. After development for over twenty years, besides the head office, Fong & Wong has established high-end professional and international design and gement service teams in Shenzhen, Shanghai and Chengdu. Fong & Wong has attracted excellent partners and designers from many countries and areas including China Mainland, Hong Kong SAR, Italy and Canada. By virtue of the design philosophy of constant innovation and pragmatic services, Fong & Wong has won a number of customers' support and trust. The cooperation customers cover both well-known Hong Kong property developers including Wharf, Sino Group, g Kong Land, South Asia Real Estate, and Newly Built Construction Property, etc. and famous mainland property developers including Zhonghai, Longfor, Country Garden, ShiMao and COFCO, China Resources Land, Greenland, China Railway Construction Real Estate Group, and OTC etc. Its design projects and works can be seen all over New York, Hong Kong, Macao, Beijing, Shanghai, Guangzhou, Shenzhen, Xiamen, Tianjin, Chengdu, Chongqing, Shenyang and many other cities and areas.

FONGWONG.HK
香港方黄

图书在版编目（CIP）数据

设计有约 2 / 方峻 著 . – 武汉 : 华中科技大学出版社 , 2016.11

ISBN 978-7-5680-2265-1

Ⅰ . ①设… Ⅱ . ①方… Ⅲ . ①室内装饰设计 Ⅳ . ① TU238

中国版本图书馆 CIP 数据核字（2016）第 238346 号

设计有约 2
Sheji Youyue 2

方峻 著

| 出版发行：华中科技大学出版社（中国·武汉） | 电话：（027）81321913 |
| 武汉市东湖新技术开发区华工科技园 | 邮编：430223 |

| 责任编辑：熊纯 | 特邀编辑：董莉婷 | 排版设计：筑美文化 |
| 责任校对：赵营涛 | 封面设计：王伟 | 责任监印：张贵君 |

印　　刷：中华商务联合印刷（广东）有限公司

开　　本：889 mm × 1194 mm　1/12

印　　张：37（上册 18 印张，下册 19 印张）

字　　数：222 千字

版　　次：2016 年 11 月第 1 版 第 1 次印刷

定　　价：558.00 元（USD 111.99）